T0209777

THE LITTLE BOOK OF BIG FUN ANIMAL FACTS

THE LITTLE BOOK OF BIG FUN ANIMAL FACTS

LOWELL R TORRES

iUniverse

THE LITTLE BOOK OF BIG FUN ANIMAL FACTS

iUniverse books may be ordered through booksellers or by contacting:

iUniverse
1663 Liberty Drive
Bloomington, IN 47403
www.iuniverse.com
1-800-Authors (1-800-288-4677)

ISBN: 978-1-6632-0253-6 (sc)
ISBN: 978-1-6632-0254-3 (e)

Print information available on the last page.

iUniverse rev. date: 06/15/2020

PREFACE

This is my fifth book with iUniverse, part of a benefit package from my days working for Author Solutions, iUnivere's parent company.

Animals, they're everywhere. And a lot of them are pretty darn cool. Here are some reasons why:

A Seychelles giant tortoise named Jonathan,
who hatched in 1832, still lives on the island of
Saint Helena in the South Atlantic Ocean.

Leatherback sea turtles can grow up to over 7 feet long and
weigh up to 1500 pounds. That's as heavy as a cow! And
they've been measured swimming at speeds up to 22 mph.

It took over 300 years for the Giant Tortoise to get a
scientific name because they taste too good. For centuries,
scientists and sailors would collect Giant Tortoises to take
back to Europe for study, but they would all eventually
get eaten. Even Charles Darwin couldn't resist, having
written "the breast-plate roasted, with the flesh on it, is
very good; and the young tortoises make excellent soup"

In 2006, Adwaitya the Aldabra giant tortoise (Geochelone Gigantes) died in Kolkata Zoo at the age of 255.

The Galapagos Tortoise can go up to a year without eating or drinking water.

Chelonaphobia is the fear of turtles.

The ancient turtle Carbonemys cofinii was as big as a small car. Its name means "coal turtle' because it was found in a Colombian coal mine.

There was once so many sea turtles in the Caribbean
Sea that sailors voyaging with Christopher Columbus
in 1492 complained of being kept awake at night by
the turtles bumping against the hulls of the ships.

The fear of birds is called ornithophobia.

Researchers taught African grey parrots how to buy food using
tokens. They paired them up and gave one parrot ten tokens
the other zero. Without prompting, the parrots with tokens
would give up some to their partners so they both could eat.

The Great Horned Owl is responsible for the iconic
"hooting" that shows up in pop culture.

The "eagle" scream often heard in pop culture
is actually the red-tailed hawk.

The Peregrine Falcon, also known as the duck hawk,
can reach speeds of over 200 mph when hunting.

Social Weaver birds build the largest nests of any
bird, big enough to house over 200 birds.

Little Penguins, known also as blue penguins or
fairy penguins in Australia, are the smallest of all
penguins, averaging just a little over one foot tall.

Scientists found that "city" birds are smarter than "country"
birds. A study showed that birds from urban environments
had better problem-solving skills then birds from rural areas.

Sumatran rhinos are the smallest living rhino, at just 3 to 5
feet tall. They're also the hairiest rhino and are more closely
related to the extinct wooly rhino than any living rhino species.

Rhino horns are made of keratin, the same protein
that makes our hair, finger and toe nails.

Rhinos have very poor eyesight, which is compensated
with long tube-shaped ears that they can swivel
in any direction to listen to faraway sounds.

The word rhinoceros comes from the Greek words rhino
and keras, which in English translates to nosehorn.

Orycteropus afer, the Aardvark, is the only species of its order.
Though once believed to be a relative of the anteater, study
has shown that it's not closely related to any living mammal.

Cats step with both their left legs at once and then
both right legs when running or walking; camels and
giraffes are the only other animals to do this.

Fromage de Chat is cheese made from cat milk, which owes its origins to pharaohs in Ancient Egypt who demanded cat cheese on their dinner tables.

Genetically the average house cat is 95% tiger.

Ben Rea loved his cat Blackie. A lot. Like, a whole lot. He loved Blackie so much that when Ben died in 1988 he left $13,000,000 (million!) to his cat.

A female cat is called a queen or a molly, a group of cats is called a clowder and a group of kittens is a kindle.

Cats usually spend 30% of their day grooming themselves.

The technical word for a cat's hairball is bezoar.

Pope Innocent VIII condemned cats as evil during the Spanish Inquisition. So many cats were destroyed the rat population exploded, helping the spread of the Black Plague.

Nyan Cat, the famous meme starring a gray cat with a Pop-Tart body shooting rainbows from its butt, was based on a real feline. Nyan was based on Marty, a cat owned by the meme's illustrator Chris Torres.

VS

The biggest dogs recorded have been over 300 pounds while the biggest domestic cat recorded was just over 45 lbs. The smallest cats and dogs both weigh about 1-2 lbs.

Adult cats will only meow for humans, while dogs will bark at anything.

A running cat can reach a top speed of 30 mph while the greyhound can run up to 45 mph.

There are 40 different recognized breeds of cat and over 200 recognized breeds of dogs

When a puppy is one-year-old it is on on the same scale of physical maturity as a 15-year-old human.

A dog's nose contains a print as unique as a human's fingerprint.

Humans have 5 million smell-detecting cells. Dogs have 220 million. A dog's sense of smell is so strong that dogs can smell emotions on humans through their sweat.

It's a myth that dogs are color blind. They can actually see color about as well as humans can at dusk.

Three dogs survived the Titanic sinking.

Paul McCartney recorded a high-pitched dog whistle at the end of the Beatles song "A Day in the Life" for his sheepdog

Spiked collars originated in Ancient Greece when shepherds would use them to protect their dogs from wolf attacks.

Dogs can understand around 250 words and gestures, the same as a two-year-old child.

A dog's loose skin allows it to shake off moisture so
well it can dry 70% of its fur in a 4 second shake.

In the 1920s the government allowed the extermination of
the grey wolf in Yellowstone and the ecosystem collapsed.
In 1995 conservationists released grey wolves into
Yellowstone again and today the area is thriving again.

That's because wolves are a keystone species, which means other animals and the land they live on all rely on that species (or kind) of animal. For example, when the wolves in Yellowstone were wiped out, the elk population grew to be huge. The extra elk meant extra eating. They ate berries that grizzly bears liked. They destroyed the willow trees that beavers liked to use to build their dams, which create marshes used by moose and otters. The reintroduction of wolves helped change the very land they roamed.

A single wolf can eat up to 20 pounds at a time.

Wolves can sprint up to almost 40 mph.

Wolves like to roam, they can travel up
to 12 miles in just one day.

Wolves can be as long as people are tall, with
males growing up to 6.5 feet long.

If the conditions are right, a wolf can hear
up to 6 miles away in the forest.

There were once so many wolves in Ireland (in the 1600s) it was nicknamed "Wolf-land."

Pandas don't have regular sleeping spots, they just fall asleep wherever they happen to be.

A grizzly bear's bite is so hard it could crush a bowling ball.

Bear cubs will stay with their mama bear for up
to 2 years before going off on their own.

Polar bears can swim up to 100 miles at a time.

Teddy bears were named after US President Theodore „ 'Teddy'
Roosevelt, who owned a small bear named Jonathan Edwards.

A bear's claws can be up to 6 inches long
and thick as a man's thumb.

Bears are considered to be one of the most intelligent land animals in North America, comparing to the higher primates.

African elephants are the largest land mammals on the planet.

Elephants live in families called herds and they're usually led by the oldest female, called the Matriarch. The males in the herds usually leave between ages 12 and 15 to live in all-male groups.

Elephants can live up to 60 years.

Elephant Appreciation Day is September 22

Elephants can get sunburned, which is why you
see elephants throwing sand on their backs and
heads. It also helps keep the bugs off.

An elephant's trunk is like a mouth, nose, and hand all in one. It can be used to pick up a penny or peel a banana.

Elephant Word Search

ELEPHANT, TRUNK, MATRIARCH, HERD, AFRICAN, ASIAN, TUSKS, PACHIDERM,

GTUPOELEPHANTLADMNEL
EFTAGEIDUDHANTRUNKSO
SOINQAAMATRIARCHBUTEL
BTNECUNKPACHIDERMOITP
ACDFQPMNZPIDUMBODNUA
BIAOABCXDRGJPOSHERDISO
ASTOSSDLKJTUENXCUPAKTD
ATHSIEYIWPOXNDOIJSOISHQ
RAVANUBAMICEWABTKAIYT
AFRICANOABHDCEAMBOHD
SKIGWHZIZOEUNSOIEHTENF

The grasshopper mouse hunts arthropods like scorpions and centipedes because they are immune to the venoms. It's famous for howling like a wolf to defend its territory: https://youtu.be/izizsAodOCk

Animal Appreciation Days

January
Adopt a Rescued Bird Month

5th - Birds

20th - Penguins

21th - Squirrels

February
Spay/Neuter Awareness

2nd - Groundhog Day

3rd - Golden Retriever

27th - Polar Bears

March
Adopt a Guinea Pig Month

13th - K9 Veterans

23rd - Puppies

April
Canine Fitness Month

14th - Dolphins

17th - Bats

29 - Guide Dogs

May
National Pet Month

14 – Chihuahuas

23 - World Turtle Day

June
Adopt-a-Cat Month

15 - Lobsters

20 - Ugliest Dog Festival

July	August
10- Kittens 31- National Mutt Day	8 – Cats 26- Dogs
September National Service Dog Month 22- Elephants 24- World Gorilla Day 26- Bunnies	**October** Adopt -a-Dog Month 8- Octopus 20- Sloths 21- Reptiles
November Pet Cancer Awareness 2- Bison 8-14 National Animal Shelter and Rescue Appreciation Week	**December** 9 – Llamas 13- Horses 14- Monkeys

There are 264 known monkey species in the world and they are classified into two groups: Old World (Africa & Asia) and New World (Americas) monkeys.

Apes like gorillas, chimpanzees, orangutans, gibbons and bonobos are not monkeys

The Mandrill is the largest monkey, weighing on average about 75 pounds. The smallest is the Pygmy Marmoset which weighs 140 grams which is about as heavy as a baseball.

Monkeys have unique fingerprints just like humans do.

A group of monkeys is called a troop or tribe.

On Yakushima Island in Japan, monkeys will groom and share food with deer in exchange for rides.

White-faced capuchin monkeys will rub their fur with Giant African Millipedes, which works like insect repellent.

A new type of monkey, the Lesula monkey, was discovered as recently as 2007, in the Congo in Africa.

The howler monkey is the loudest monkey in the world. It can be heard three miles away.

The scientific name for a gorilla is Gorilla gorilla gorilla.

A lion's roar can be heard 5 miles away.

Despite being "King of the Jungle" only one population of Lions live in a forest in India. All other lions live in savannah and grasslands.

A lion can run as fast as 50 mph on the hunt and leap over 30 feet.

A lion might sleep up to 20 hours every day.

Aslan is the Turkish and Mongolian word for lion, so "Aslan the lion" (from the Chronicles of Narnia) translates to "Lion the lion."

Giant river otters live along the Amazon River basin and can grow 6 feet long. They live in groups of up to 20 called a romp. Nicknamed "wolves of the river" they have been known to hunt other big predators like anaconda, piranha, and caiman.

In Singapore lives the Bishan family of otters, which rules the nearby rivers and will often rove up and down the waterways like gangs, to fight other families and take over their territory.

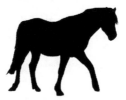

Horses can run just a couple hours after being born.

Horses hooves are made from the same protein that makes up human hair and fingernails.

Horses can sleep laying down or standing up.

Horse's eyes are on the side of their head, which allows them to see nearly 360 degrees around them.

The only truly wild horses left on the planet are the Przewalski horses of Mongolia. Other populations around the world are considered feral.

A horse will often point their ears where they're looking. If the ears are pointed in different directions, the horse is looking at two different things at the same time.

A horse's ear has 16 muscles in it, which lets
them turn their ears 180 degrees.

A horse can produce up to 10 gallons of saliva a day.

The fastest a horse has ever been recorded is 55mph

A red ribbon on a horse's tail means it kicks.

While other mammals are capable of gliding,
bats are the only mammal that can fly.

Just one bat cat eat over 500 bugs in one hour.
That's like eating 15 pizzas.

There are more than 1,100 bat species in the world,
making up one quarter of all mammals.

The bumblebee bat is the smallest bat, weighing
less than a penny. The largest is the giant golden-
crowned flying fox with a wingspan of 5 to 6 feet.

Saltwater crocodiles are the largest crocodiles in the world, growing up to 23 feet long and 2200 pounds.

There are more than 4500 species of crab, ranging in size from 12 feet across to just half an inch big.

Ostriches are the largest birds in the world and also the
fastest on land, running over 40 miles per hour.

Ostriches don't have teeth, so they will eat pebbles
or sand to help break their food down.

A zebra's stripes are as unique as a person's fingerprints
and can be scanned by scientists to tell individuals apart.

Camels have two rows of eyelashes and three
eyelids to help keep sand out of their eyes.

Gazelle can gather in herds of 700 and can run
up to 60 mph to get away from predators

The strike of an eagle can be up to two
times harder than a rifle shot.

Badgers will fight off much larger animals like bears and
wolves, and have been seen hunting with coyotes.

An allosaurus never stopped growing in
the 25 or so years they lived.

Velociraptor in Latin means "swift seizer" or "speedy thief"
because velociraptors, contrary to what the movie Jurassic
Park showed, were scavengers about the size of a turkey.

Carnotaurus is the only carnivorous
dinosaur with horns on its head.

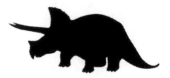

Triceratops means "three-horned face" in greek but the horn on the nose was actually made of keratin, the same material as our fingernails.

Stegosaurus was among the dumbest of the dinosaurs. While it was the size and strong defensive plates were intimidating, its brain was the size of a walnut.

Duck-billed dinosaurs, or hadrosaurs, lived as far north as Alaska.

Apatosaurus, also known as brontosaurus, weighed 50,000 pounds.

Pachycephalosaurus had a skull up to 10 inches thick

Argentinosaurus was the longest and heaviest dinosaur, measuring over 120 feet and 77 tons. That's as long as four fire engines and as much as 17 elephants put together.

The biggest dinosaur found is Argentinosaurus, which weighed up to 200,000 pounds and over 100 feet long

Pterodactyls weren't dinosaurs, but flying reptiles that happened to live during the age of dinosaurs.

Plesiosaur was also a reptile

The Best Websites About Animals

https://kids.sandiegozoo.org/
https://www.nationalgeographic.com/animals/facts-pictures/
https://www.all-birds.com/
https://a-z-animals.com/
https://www.animalfactsencyclopedia.com/

Printed in the United States
By Bookmasters